SOYA
JENNY
BOTSFORD

Soya

First cultivated thousands of years ago in ancient China, the soya bean is fast becoming the most important crop grown in the world today. Protein-packed and full of oil, this versatile little bean is used in a surprising number of foods for both animal and human consumption, and also plays an important part in many industrial processes.

This book tells the story of soya ... how its cultivation spread from East to West; the discovery of its high protein content; where and how it is grown today; its importance on the world market; and its role in the future.

Jenny Botsford is a public relations consultant and also represents the Vegetable Protein Association.

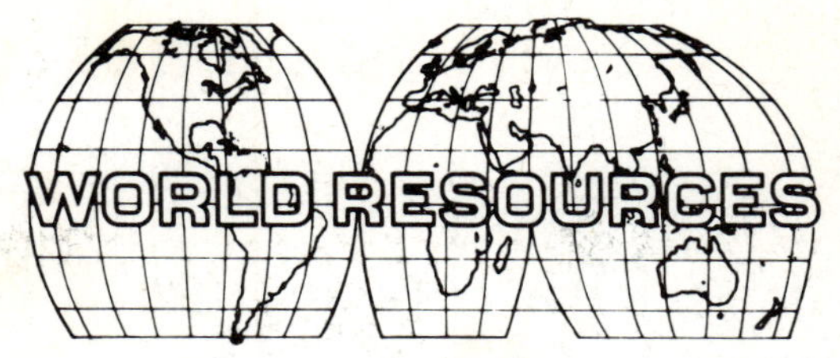

SOYA

Caister Middle School
Kingston Avenue
Caister-on-Sea

JENNY BOTSFORD

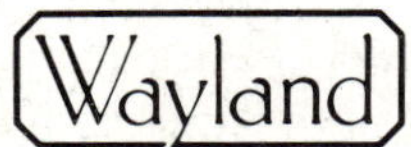

ISBN 0 85340 676 6
© Copyright 1980 Wayland Publishers Limited
First published in 1980 by
Wayland Publishers Limited
49 Lansdowne Place, Hove
East Sussex BN3 1HF, England

Phototypeset by Computacomp (UK) Ltd
Fort William, Scotland
Printed and bound in Great Britain at
The Pitman Press, Bath

World Resources Series

Timber	Copper
Rubber	Rice
Iron and Steel	Diamonds
Sugar	Alternative Energy
Seafood	Paper
Oil	Soya
Gas	Aluminium
Coal	Glass
Cocoa, Tea and Coffee	Plastics
Grain	Leather
Meat	Dairy Produce
Nuclear Fuel	Vegetable Oils
Gold and Silver	Maize
Tobacco	Cotton
Wool	Salt

Frontispiece *Children watching the soya bean harvest in the U.S.A.*

Contents

Above   *Giant storage elevators loom over a field of growing soya.*

Above   *Soya beans can be cooked in a variety of dishes.*

6

What is soya?

The gold that grows

The bean – one of the oldest vegetables known to man – has long been recognized as a wholesome source of body-building protein. But it is only relatively recently that the true value of the soya bean has been fully realized. In terms of protein, mineral, oil and vitamin content soya is 'King Bean'.

Soya beans are the little round or oval seeds of the soya plant. Like peas and broad beans, which belong to the same *legume* family, soya beans grow in pods on leafy plants. In Western countries we are more familiar with peas and broad beans. But if you visit a home in China, Japan or Indonesia, you will probably eat soya beans in one form or another, for soya is an important part of the Oriental diet.

You may think soya is a new discovery. But it is in fact one of the oldest crops grown by man and has been valued in the Orient as food and for medicinal purposes for many thousands of

Above *The beans of the soya plant grow in hairy pods which burst open when the beans are ripe.*

years. Today soya is becoming the most important crop in the world. It is the most efficient and least costly source of protein as well as an important source of oil. Many people see it as a part-solution to the problem of the world's starving millions.

But soya is not only an important source of food, it also has hundreds of uses in industry, playing a part in the manufacture of a huge range of products from explosives and insecticides to paint and plastics. There is little doubt that this 'miracle crop' will play an increasingly important role in the twentieth century as scientists find yet more applications for it.

Above   *The ancient Chinese recognized the nutritional value of the soya bean long before it aroused any interest in the Western world.*

The beginnings of soya

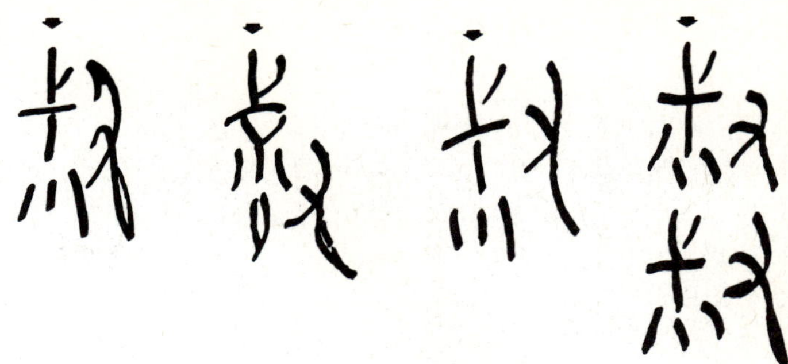

Soya in ancient China

According to Chinese tradition there once lived an ancient Chinese ruler called Shen Nung. Before his reign the people were nomads. They wandered from place to place gathering wild plants and berries, and killing animals when they could for food. The soya plant was one of the wild plants they gathered.

Shen Nung, known in China as the Father of Agriculture and Medicine, taught the people how to plough their land and sow seed. This meant they could settle in one place and grow food to store against famine. He also made medicines from plants and in 2838 B.C. wrote a book about his herbal remedies. In his book he gave instructions for making over 300 medicines, which were all prepared from soya beans. This was thought to be the first description of the soya bean.

But scholars have now found earlier mentions of soya. The ancient Chinese character for soya has been found in the Book of Odes, which was probably written as long ago as the eleventh century B.C. Some historians now believe that the writings of Shen Nung are fabrications of earlier historians and that the tale of the Emperor himself is a myth. However, the writing in the Book of Odes shows that people must have been growing soya in China long before the time of Shen Nung, if he existed, and in any case thousands of years before the birth of Western civilization.

The ancient Chinese realized the value of the soya bean and classed it as one of the five sacred 'grains', together with rice, wheat, barley and millet. These crops were all considered vital.

The soya bean has stood the test of time and is becoming as important to twentieth-century man as it was to the Chinese 5000 years ago.

From East to West

Soya was probably first cultivated in north-east China. From there it spread to the rest of the Orient, as traders took soya beans with them on their sea voyages. Soon soya was growing in south China, Japan, Korea and South-east Asia, where it adapted well to different soils and climate. But for centuries soya was only grown in the East.

However, by the seventeenth century explorers and missionaries had brought soya seeds back with them from their travels. Attempts to grow soya in England, Germany, France and Hungary met with little success. Records show that soya beans were grown as early as 1790 in the Royal Botanic Gardens at Kew, England. However, the crop was not well suited to the cold English climate and was not grown commercially.

Soya was first taken to America in 1804 as ballast in a ship. When cargo had been unloaded in the Orient, the ship would be too light to return to America safely. If there was no cargo to take back, something which was local and cheaply available, such as soya, was taken on as ballast to stabilize the ship.

Settlers planted the soya beans in America, where the plant grew very well. The Perry expedition to Japan in 1854 brought back different varieties of soya beans which proved to be well suited to the American climate and soil. The soya bean was then known as the 'Japan pea'. In 1882 the Carolina Experimental Agriculture Station produced a bean known as the Mammoth Yellow, which was a leading variety for many years. The U.S. Department of Agriculture quickly realized the value of the soya bean to American farmers, and in 1890 started a research project on the use of soya to increase soil fertility and as silage. Soya's two main products – oil and meal – had not yet been discovered.

Right *Soya first arrived in the U.S.A. as ballast on a ship. Today soya is one of the U.S.A.'s main items of trade.*

The start of soya processing

George Washington Carver, an American ex-slave, began research in 1904, which was to reveal the secrets of the soya bean to the Western world. He discovered the exceptionally high protein content of the bean. We now know that the soya bean is the richest source of protein, containing 40 per cent protein (more than twice the protein found in meat or fish), 20 per cent vegetable oil, plus useful amounts of vitamins, minerals and trace elements, all essential for healthy living.

Until then the soya bean had only been used as a forage crop and for green manure, which was ploughed back into the land. But gradually chemists and manufacturers came to recognize the value of the soya bean itself, and the great soya processing industry was born. The first commercial processing of soya bean seed to obtain oil and cake took place in 1911, when a small hydraulic press-mill processed a shipment of beans from Manchuria. Oil and protein meal from American soya beans were first produced in 1915 in a cottonseed mill in North Carolina. During the 1920s Dr. W. J. Morse, who is often called the Father of the American Soybean Industry, travelled to China and brought back thousands of different varieties of the bean. Until then the amount of soya grown commercially in the U.S.A. was of little

12

significance. He tested the Chinese plants to see which grew best. This led to the discovery of varieties which could be grown profitably.

The stage was set for the growth of the America soya bean industry. The crop gradually became recognized as an important source of oil. By 1935 the acreage of beans grown for processing equalled that of beans grown for forage, and then quickly overtook it. The rapid expansion of the industry was partly due to the world shortage of edible oils and also to the Second World War, which put traditional sources of protein in short supply. Research began in the U.S.A. and England into ways of using the protein-rich soya meal for human food, but it was not until the late 1950s that the first edible soya protein plant was cultivated in Chicago.

Opposite *Interest in the protein content of soya was aroused during the Second World War when traditional sources of protein were in short supply.*

Growing soya

The soya plant

Soya beans grow in pods on hardy, upright plants. The soya plant has a strong tap (or main) root with many branches that go deep into the soil. This deep root system enables the soya plant to withstand drought, as the roots can reach water well below the surface. The bushy plant usually grows to a height of 90–120 cm (3–4 ft), although some varieties can grow as tall as 150 cm (5 ft), if the land is irrigated.

The leaves are usually dark green with three lobes (or parts). As the soya pod ripens the leaves turn yellow and drop by the time the beans are ready to be harvested. Small pink, purple or white flowers appear where the leaf joins the stem. The soya plant is self-fertilizing as the flowers contain both male and female parts. The leaves, stems and pods are all covered with short, fine hairs. The bean pods, which vary in colour from light yellow to brown, grey and black, grow close together on the upright stem.

14

Above *A close-up view of the soya plant showing the hairy pods and upright stem.*

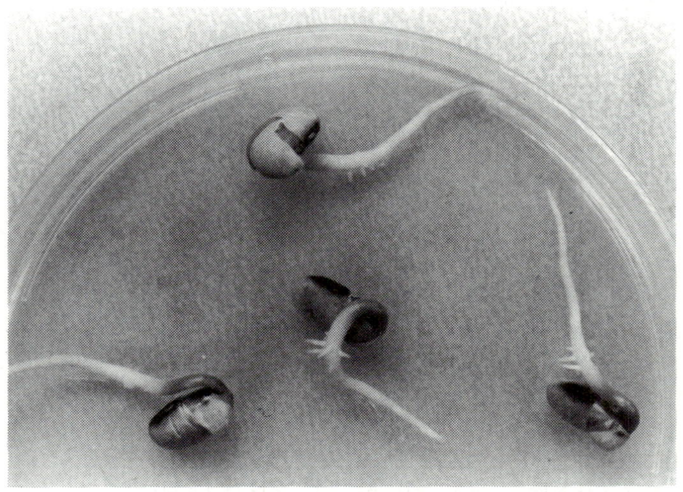

Above *As soon as the seeds germinate the strong tap root emerges.*

Inside the hairy pods are one to four small round or oval seeds about the same size as a pea. The beans may be yellow, green, brown, black or mottled, although the yellow variety is the most popular for crushing and processing as it contains the most oil. Each bean has a saddle-like patch of black, brown or grey down one side. This is the hilum (or scar) where the bean was joined to the pod.

There are thousands of varieties of soya, each of which is suited to different growing conditions. But as yet there are no true hybrid soya plants, although researchers are currently working on this problem. It is hoped that by crossing two strong parent plants a new and even stronger variety of bean will be produced.

Planting

The soya plant is an annual, which means it lives for just one year. New seeds must be planted every year. Soya likes a warm climate with plenty of sun. Different varieties have been bred to grow in a range of climates from tropical to temperate. As long as the summers are very warm, soya will grow well.

The growing season is measured from the date the seed is planted until the plant is ready to harvest. This varies from about 180 days in northern, temperate regions to 75 days in warmer, more tropical areas. During the growing season the crop needs a good supply of rain. Ideally rainy periods should alternate with sunny spells. The seeds should be planted during the rainy season with the plants maturing when it is dry.

Soya beans will grow successfully on most types of soil, although they prefer a well-drained, fertile, loamy soil. Loam is a mixture of rich vegetable mould with clay and sand. Good crops of soya can also be grown on drained swamplands. Areas where maize, wheat or

Right *On the vast soya bean plantations of the U.S.A. seeds are planted by machine.*

16

Above   *Wheat straw mulch is left on the ground between the rows of plants to protect the soil from erosion.*

groundnuts grow well are usually suitable for soya bean cultivation. Soya is usually grown in rotation with a number of other crops, especially maize.

If the land is lacking in nourishment, lime and fertilizer are applied before planting. Like

Above *Research is currently being conducted to discover whether soya can be grown in dry areas by irrigating the land.*

other legumes, soya beans obtain part of the nitrogen they require from the soil through the action of special bacteria living on the roots of the plant. This helps the nodules (or root branches) to grow.

Before the seeds are planted the soil is tilled to make a smooth seed bed. The seeds, which must be clean, healthy and recently harvested, are planted from early spring to midsummer. In the U.S.A. planting takes place in May and early June. Sometimes soya is planted as a second crop following potatoes or early vegetables, or to replace a crop which has failed. The beans are usually planted in rows about one metre apart, so that machinery can get between the rows.

Weeds must be controlled with a rotary hoe, spike-tooth harrow or heavy weeder to prevent them choking the young plant.

In Oriental countries most soya is still grown in the same way it has been for centuries. The crop is grown in rotation with grain – soya one year, grain the next and so on. This helps to keep the goodness in the soil. There is a very wide variety of soya plants, each suiting the local conditions where it is grown. The soya seeds are sown by hand on manured land each spring. At the end of the summer, when the plants have grown and the bean pods are fully ripe, they are gathered by hand.

Above   *Before the seeds are planted, the soil must be tilled to break up any clods of earth.*

Above   *The young plants must be protected from stronger weeds by regular hoeing.*

19

Soya's enemies

Farmers take great care to protect their crops against pests and disease, inspecting the young plants regularly so that any problems can be tackled immediately.

Insects which may attack the soya bean crop include the velvet bean caterpillar, grasshopper, blister beetle, stink bug, leaf hopper and many others. These insects rarely spoil the whole crop, but hungry rabbits and wood chucks can cause a lot of damage. Caterpillars or beetles may eat their way through leaves, stems, buds, flowers or pods. Little insects sometimes cluster under the leaves making them change colour and eventually causing the whole plant to shrivel up or grow very slowly. This drying up can also be due to large insects which suck the plant tissues.

Living in the soil are microscopically small eel-like worms called nematodes. If the same field is used for growing beans year after year the nematodes increase. If they are present in large numbers in the soil, knots and swellings form on the soya bean roots, causing the plant to turn yellow and stop growing. One kind of

Left *Soil samples are tested to determine the level of cyst nematode present in the soil.*

Above   Smartweed can quickly strangle weaker soya plants, if it is not promptly weeded out.

Above   No sooner has this young soya plant pushed its first pair of leaves above ground than it is attacked by cocklebur.

Above *By studying diseases which attack soya, scientists hope to breed disease-resistent varieties.*

Above *The meadow grasshopper sucks the tissues of the soya plant, eventually causing it to wither and die.*

nematode is so dangerous to soya bean production that strict quarantine is imposed in the U.S.A. wherever it is found.

Diseases can be carried to plants by insects. It

is in fact very rare to find a completely disease-free field. Some diseases can cause a whole crop to die, while others are less serious. The names of some of these diseases are very descriptive. They include fungus diseases such as frog eye, brown spot, target spot, stem and root rots, and virus diseases such as soya bean mosaic and bud blight. Agricultural scientists have bred special strains of soya plants which can resist some of these diseases, and seeds can also be specially treated against disease. New ways to conquer pests and disease are continually being studied. This is one way in which crop yields can be increased.

Above   *Regular application of herbicide is one way in which soya plants can be protected from weeds.*

Harvesting

The American soya bean crop is ready for harvesting during September and October, depending on the variety of bean, length of growing season and climate. As the crop prepares itself for harvesting, the leaves turn yellow and drop off, and the pods and stems dry out.

The best time for harvesting is when the branches and leaves begin to shrivel, but before the pods have shattered and burst open. The stems should still be moist and pliable. It is very important to harvest the crop at the right time to get the best yield. If harvesting is done too soon, before the pods are mature, the yield will be lower and of inferior quality. If the crop is mainly required for hay, silage and animal fodder, it is cut a little earlier before most of the leaves have dropped.

In the U.S.A. all harvesting is done by huge combine harvesters, or harvester threshers. On small-scale plantations in other countries the whole plant is pulled out of the earth by hand and left to dry in the field. The farmers then choose the healthiest plants, which are kept

Above *Most of the soya grown in the developed world is harvested by huge combine harvesters.*

Above   *Over 70 million acres of land in the U.S.A. is devoted to the cultivation of soya.*

whole and unthreshed through the winter. These will provide seeds for next year's crop. By keeping the plant whole the seeds are protected in their pods from disease and insects. If a large number of seeds is required and the unthreshed plant would take up too much storage room, the best plants are threshed and the seeds are stored in airtight containers.

Below *The use of sophisticated machinery enables the farmer to harvest his fields quickly and efficiently.*

Above   *A harvester thresher machine unloads the threshed beans into a wagon.*

Above   *In poorer parts of the world soya is still threshed by hand.*

Threshing and storage

Threshing, or thrashing, means just that. The plants are beaten to separate the bean from the other parts of the plant. On a small scale threshing can be done by hand with a stick, or flail, on a threshing floor, either indoors or outside. This is how threshing was done centuries ago, and indeed still is in many poorer parts of the world.

To thresh larger quantities of beans a seed winnower can be attached to a tractor. This separates the bean pods from the rest of the plant. The whole plants are spread out on a hard threshing floor first. On large-scale plantations thresher harvesters are usually used.

Below *Many soya bean farmers store their crop on the farm in giant silos.*

Above *Many farmers store their beans in giant silos shared by local farmers.*

These machines thresh the soya as they cut it down.

When the beans have been threshed, they have to be stored. The beans must not be too moist or there is a danger that they may rot. Usually the soya is tested while it is growing to determine moisture content. When the beans have a moisture content of 13 per cent or less the farmer knows they can be safely stored. If they are too moist when harvested, they must be dried out. Natural drying is the cheapest way, but artificial drying may be necessary. On a small scale the beans can be left in the sun to dry. Some of the soya beans are stored on the farms in bins or large elevators. Others are taken to silos, which are central to all the farms in the area, or straight to the processing plant. By storing the beans in large silos the manufacturers have their raw material close to hand and are unaffected by day-to-day price changes.

If the soya is stored in airtight containers, and is not moist, it will keep very well for three years or more.

30

Above   *The soya beans must not be too moist when they are stored or they will rot.*

Above   *Loaded trucks transport harvested beans from the field to storage elevators.*

Above   *The various stages in the processing of soya are controlled by sophisticated electronic equipment.*

At the processing plant

Cleaning and cracking

When the soya beans arrive at the processing plant, they are cleaned to remove any foreign objects, which may have become mixed with the crop during the harvest. This is done by selective screening and air cleaning. The beans are put through a special sort of sieve which lets the beans through, but keeps back anything larger. Electromagnets pick out any metal objects. The beans may then be passed through a blowing chamber, where a current of air is blown through them, and anything heavy such as grit falls to the bottom. The beans are then stored ready for further processing.

Before the dried beans are processed, they must be screened once again. All damaged or broken ones are removed as well as any other foreign matter which may have escaped the first screening.

The beans then pass through corrugated rollers, which can be set so that the beans are

Above *One of the best ways of transporting soya from the farm to the factory is by barge like this one on the River Mississsippi.*

cracked into quarters or eighths. At the same time the hull (or seed coat) is loosened from the cotyledon (the fleshy part of the bean). The hulls are not wasted, but are treated with a heat process, and milled or ground to make bran. Bran is an important dietary fibre which aids digestion, and is used in crispbreads and breakfast cereals. It is also used in animal meals, or sold as mill feed, which is used to add fibre to certain animal feeds. For example, molasses or dark treacle contains useful minerals, but would be very difficult to feed straight to animals. By mixing it with mill feed, it can be made into pellets which can easily be fed to animals.

Extracting the oil

After the cracked beans have been dehulled, some are toasted, cooled and then ground up coarsely to produce full fat soya flour. This is used in the baking industry, and in some animal feeds. Soya bean flour is particularly valuable in developing countries where special diets have been created using soya. It is one of the best-value foods available to man in terms of protein, minerals and vitamins.

After dehulling, the soya beans are flaked, which makes it easier to extract the oil. The beans are usually heated and steamed in giant 'kettles' to make them soft and ready for the flaking process. They are then squeezed between huge rollers, which produce very thin flakes.

In the past the oil was squeezed out of the beans by an expeller, which worked rather like a giant mincing machine. Now a new method has been introduced which extracts more oil out of the soya. This is the solvent extraction process, which dissolves the oil out of the beans.

The flakes go to an extraction tower, where they are moved around inside on wire trays. Jets of solvent, usually hexane, are sprayed onto the flakes, which dissolve the oil. The hexane

Right *One of the main uses of crude soya bean oil is in the manufacture of salad dressings.*

washes the oil out through the wire trays, making a new mixture which is part oil/part hexane. This drains down to a collecting tank. Meanwhile the defatted flakes go on for further processing. The oil/solvent mixture is heated to a temperature where the hexane turns to a gas or vapour. This vapour can then be collected, condensed, and turned back into liquid hexane which can be re-used many times. The oil which is left behind is crude soya bean oil. When refined this oil is useful in the manufacture of margarines, salad oils and cooking oils.

Above   *Soya bean flour, meal and oil are three by-products of the soya bean.*

Above   *Cottonseed oil is one of soya's main competitors in the vegetable oil market.*

Refining vegetable oil

Before crude soya oil can be used in the manufacture of food products, it must be refined to make it edible. This process involves a number of steps. In modern plants the whole process is controlled by computers.

First, the 'gums' in the crude oil are removed. The oil is poured into huge tanks, where it is mixed with water or a weak salt solution. As you will know if you have tried to mix oil and water, the oil stays on top. The gums and other impurities also collect in the watery layer, which is then run off leaving the oil.

The gums are then dried. The dried gums (known as lecithin) can be purified to make them edible. Lecithin is an emulsifier – a substance which allows water and oil to mix without separating. It is used in many industrial processes such as the manufacture of rubber, pottery, textiles, cosmetics and a wide range of foods including margarine, shortening, ice cream and chocolate.

The next stage is to remove certain fatty acids which would make the oil rancid or 'off'. This is done in the same large tanks which were previously used for the degumming operation. The oil is heated and an alkali solution is sprinkled on it and stirred with big paddles for about 30 minutes. The fatty acids combine with the alkali to form a thick soap solution which

can be separated from the oil. This solution settles to the bottom of the tank, and is later sold to the soap industry. The oil is then washed and dried under vacuum to remove all remaining traces of soap.

By this time the oil has become slightly lightened in colour. It is further bleached with a special kind of Fuller's earth, which is added to the oil in the tank while it is still under vacuum. The remaining colour pigments are absorbed by the Fuller's earth, and the clear oil is then pumped out of the bottom of the tank.

Soya oil has recently become the leading vegetable oil in the world. But other oils are also used to produce margarine and cooking fats. In the U.S.A. soya and cottonseed are mainly used because they are readily available. In other parts of the world where oils have to be imported, soya, groundnut, cottonseed, coconut, sunflower and palm are all used.

Blue
Band

Hydrogenation

Hydrogenation is a very important process which makes oils go solid. Before hydrogenation was discovered early in the twentieth century, only animal fats could be used to make margarine, but now a whole range of oils can be used.

Fats and oils have the same basic structure. Those that are liquid at room temperature are known as oils and those that are solid are called fats. But you know that if you heat a fat it will run. Unsaturated fats contain less hydrogen than saturated fats, but they melt at lower temperatures and are more likely to go 'off'.

Hydrogenation is a process which adds hydrogen to the fat to make it less liquid and more stable. This is done by adding another substance – called a catalyst – to trigger off a reaction. A minute amount of finely powdered nickel acts as the catalyst to hydrogenate soya oil.

The oil is treated with hydrogen gas at high temperature under pressure in an enclosed tank in the presence of the nickel catalyst. The nickel is then filtered out. The process can be controlled and stopped at any point. Lightly hydrogenated oil will stay in a liquid form, but it will be stable. If the process is continued the oil will go solid at room temperature and can be used for margarine and shortenings.

The higher melting parts of edible oils are then removed. This process stops them going cloudy when refrigerated and is called winterization.

The final process – deodorization – removes the smell. High pressure steam is blown through the hot oil while it is still under vacuum. After five hours the oil is completely bland and does not taste or smell.

Opposite *Margarine is now a popular alternative to butter.*

Making margarine

Margarine is a relatively new food, which was invented just over a hundred years ago by a Frenchman called Hippolyte Mège-Mouriès. He had entered a competition promoted by Napoleon III, Emperor or France, to produce a new type of butter which would be cheaper to produce and would keep better.

However, the margarine we know today is very different from that invented by Monsieur Mège-Mouriès. He used animal fats, but nowadays vegetable oils – including soya bean, cottonseed, peanut and corn oils – can also be used. Recently doctors have warned that too much animal fat in the diet can lead to heart disease, so special types of margarine made with vegetable oils are becoming increasingly popular.

Margarine is made by blending fats or oils with other ingredients, normally milk products. Soya bean oil is one of the main vegetable oils used in the manufacture of margarine in the U.S.A. Emulsifiers which stop the liquid and fat from separating, flavour, colour and vitamins A and D are added to the refined oil to match the vitamin content of butter. The oil is then mixed with a milk mixture. It is chilled in a votator to

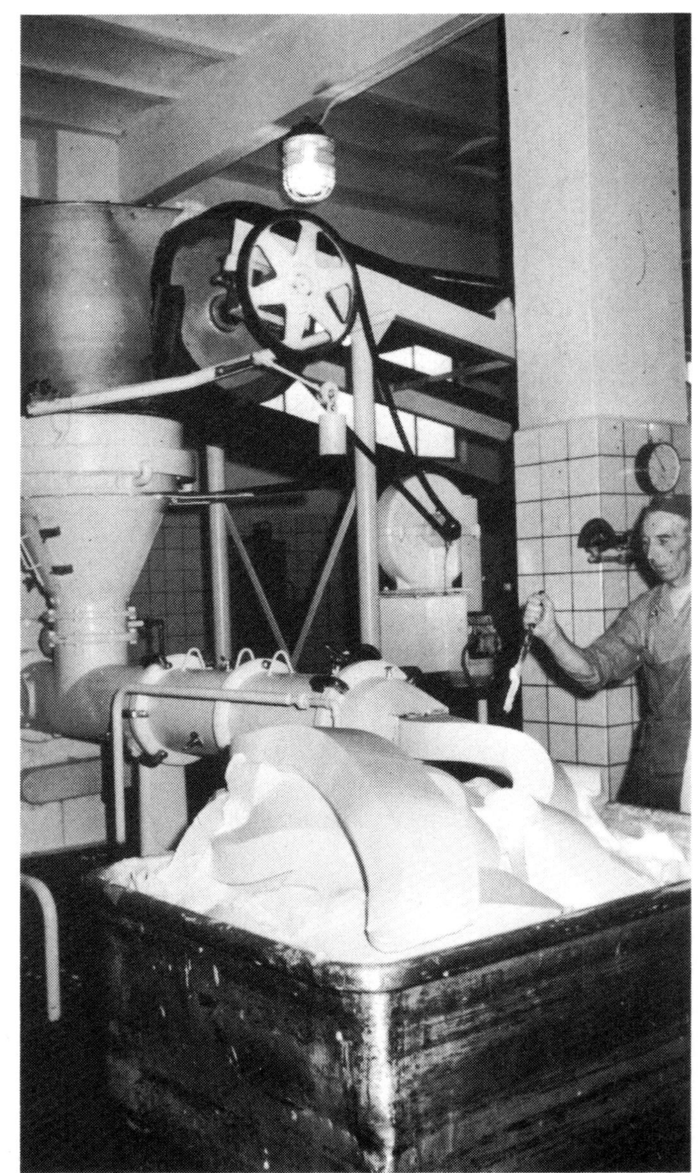

Right *Soya bean oil has to compete with other vegetable oils, such as corn, coconut and groundnut, which can all be used to make margarine.*

Above *Margarine made with vegetable oils is becoming increasingly popular as people become aware of the dangers of too much animal fat in the diet.*

ensure that it is blended evenly and the right texture is obtained. It is then packed into tubs or cut in blocks and wrapped in a special covering to stop it deteriorating.

Margarine can be made in different consistencies to suit different needs. Table margarine must spread easily, taste good without being fatty, fry without smoking, and cream well when used for baking.

Margarine is now used all over the world, with Asia, Africa, South America and Japan producing one third of the world's total supply.

Cooking fats are blends of oils and fats chilled and textured in the same way as margarine. Soya oil for cooking and salad dressing is bottled after light hydrogenation to make it keep well. Mayonnaise, salad dressing and sandwich spreads can all be made from soya oil with the addition of other ingredients. Soya oil is even used in the manufacture of some ice creams.

Animal feed

When the soya oil has been extracted from the soya flakes, defatted flakes rich in protein are left. About 85 per cent of these are made into soya bean cake or meal for feeding to livestock and poultry.

It is important to feed animals a balanced diet. There is no point in giving an animal large quantities of expensive food unless it puts on weight of the right kind and fetches a higher price in the cattle market. Soya bean meal is the most important of all the high-protein animal feeds.

Animal feeding efficiency has been increasing steadily. For every 45 kg (100 lb) of feed eaten, cattle now gain 7 kg (16 lb) in weight compared with 4 kg (8 lb) 40 years ago. Milk production has also doubled. This increase has coincided with more soya bean meal being fed to animals.

The public now demand lean meat. In the 1900s pigs produced 2 kg (4 lb) of pork to 0·5 kg (1 lb) of lard. By the 1970s the proportion had increased to 7:1. This higher ratio of meat to fat has been achieved by balanced protein diets, together with better breeding, and more efficient management.

Processed soya bean meal is now the chief source of protein fed to poultry. Rations are carefully balanced to achieve the highest rate of growth or egg production for the least cost.

Above *Most of the defatted flakes which remain after the oil has been extracted are made into meal.*

For some livestock feeding soya oil, full fat soya flakes, vitamins, and other ingredients are also added to the feed. Scientists are constantly experimenting with a whole range of animal foodstuffs to obtain maximum production.

Soya is even used in soya milk feeds for young animals when they have been orphaned. Soya bean flakes, soya bean pellets, fish food, dairy feeds, bee food and pet food are just some of the animals foodstuffs made from soya.

Above   *Protein-rich soya bean meal is fed to animals as part of a balanced diet.*

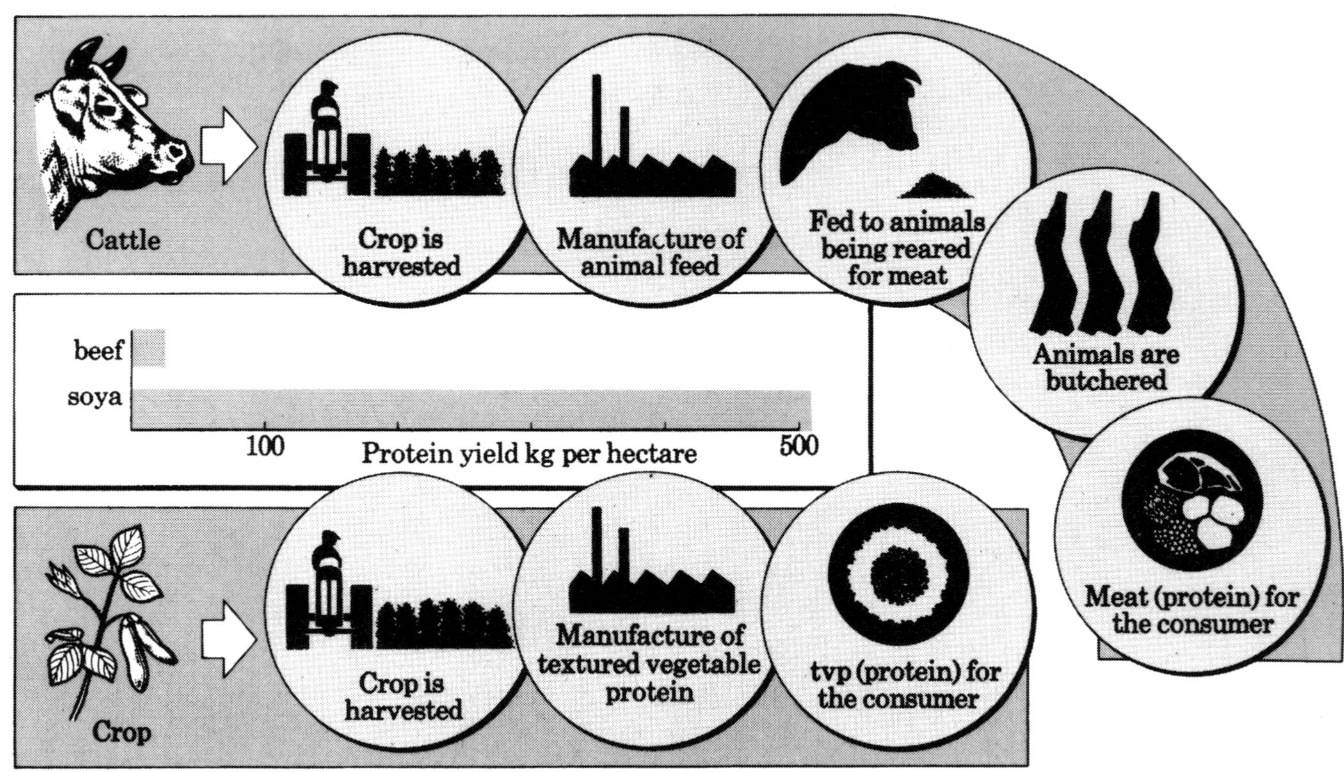

Above *This diagram shows that it costs much less to produce vegetable protein than it does to produce an equivalent amount of animal protein.*

Soya as food

Above *To many people in the West meat and two vegetables is a standard meal, but is this the best way of obtaining a balanced diet?*

Animal or vegetable?

However efficiently animals are reared, it is still not as economical a way of producing food as growing crops. A field of vegetable crops will produce far more protein per year than if it were used for raising beef cattle, poultry or pigs. The same quantity of vegetable protein costs much less to produce than animal protein. Soya has the highest protein yield of all crops. There are, of course, types of land where crops cannot be grown, but where animals can be raised. This does not alter the total picture.

The ideal diet contains both animal and vegetable protein. Most of our animal protein comes from specially reared animals. As we have seen, animals eat vegetable protein food which is then converted into animal protein. The amount of food an animal needs to produce a certain quantity of meat is called the protein conversion ratio. Animals are not very efficient protein converters. They need to eat a large amount of vegetable protein to convert it into animal protein or meat. A pig needs 5 kg (11 lb) of vegetable protein to produce 1 kg (2 lb) of meat.

Some people believe that we should make better use of the world's resources. Instead of feeding protein-rich vegetable food, such as soya, to animals, we could process more of it which could be eaten directly by human beings.

In the richer countries of the world people eat more protein than they actually need, whereas in other parts of the world such as parts of China, India, Pakistan, the Middle East, West Indies and South America, millions of people are undernourished. There is in fact no overall shortage of food in the world. If it were evenly distributed, there would be enough for everyone and even a surplus of protein.

'You are what you eat'

The food we eat supplies us with all the different materials we need to build our bodies, repair worn-out tissues, and provide energy. We need a rich supply of protein, vitamins and minerals to protect the body against disease. The soya bean contains all these ingredients.

Protein is essential to counteract daily wear and tear on the body and maintain good health. We can obtain protein from both animal and vegetable sources. Animal protein comes from meat, chicken, fish, eggs, milk and cheese. The richest source of vegetable protein is soya which contains nearly 40 per cent protein.

All living things are made up of millions of tiny cells. If you look at your skin or a drop of blood under a high-powered microscope you will see these tiny cells, which are constantly dying off and being renewed. For this they need a constant supply of protein. Old protein in the cells is replaced by the protein we eat.

Proteins are made up of chains of other smaller units called amino acids. There are 22 known amino acids, eight of which are essential to us because the body cannot manufacture them. The soya bean is one of the few foods which contains all eight essential amino acids. As such it is considered a complete protein food.

As well as protein, we also need carbohydrates, fats and vitamins. We obtain

46

Above *Fish is one of our main sources of animal protein.*

PROTEIN YIELD

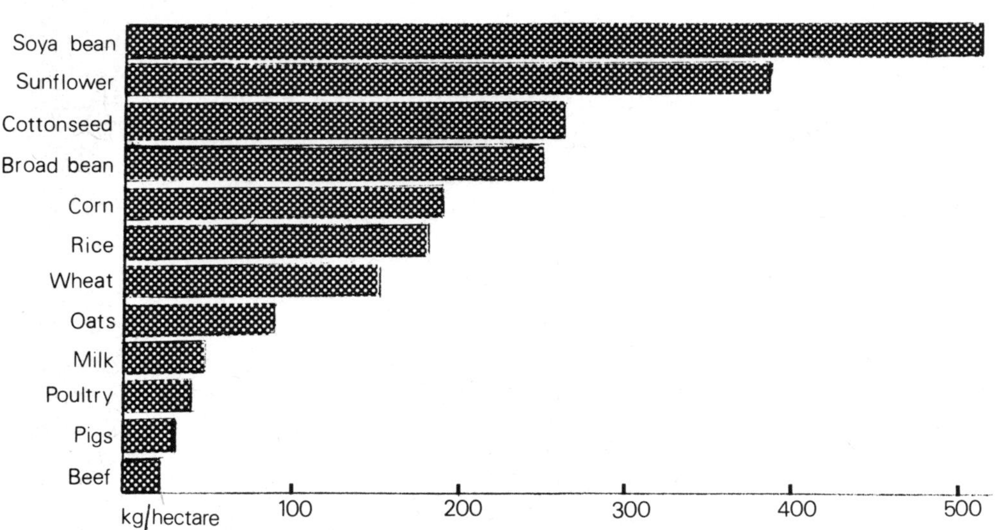

carbohydrates from starchy foods, such as potatoes, cereals, bread, rice and pasta. Fats provide us with a concentrated form of energy and add to the feeling of fullness after we have eaten. About one fifth of the soya bean is composed of fat. We also eat fats in butter, margarine, cooking fats, and oils in milk, cream, cheese, meat and fish.

An adequate supply of vitamins is also essential for good health. There are about 20 known vitamins. Soya bean sprouts are rich in vitamins A, B and C. Vitamin A guards us against infectious disease, promotes healthy skin, and also helps us see in dim light. The B group of vitamins, including thiamine, riboflavin and niacin, are essential for keeping nerves in good condition and releasing the energy from the food we eat. Vitamin C is essential for the prevention of scurvy and the maintenance of healthy connective tissues.

Soya beans also contain important minerals and trace elements. The two most important minerals present in soya are iron, which gives vitality and energy, and calcium, which is essential for healthy bones and teeth.

Soya and the food industry

Extracting the oil from soya beans and using the meal for protein-rich animal feed are not the only ways in which soya is used for food. You will be surprised to learn just how many everyday foods contain soya in one form or another because it makes them taste better, keep longer or improves them.

Full fat flour is made by grinding the fleshy part of the beans before the oil is extracted. This has been used in the baking industry for many years. A small amount is added to bread,

Below *Efforts to produce a new, high-protein food have resulted in a five-step process to produce soya bean flour without losing any of the nutritious oil.*

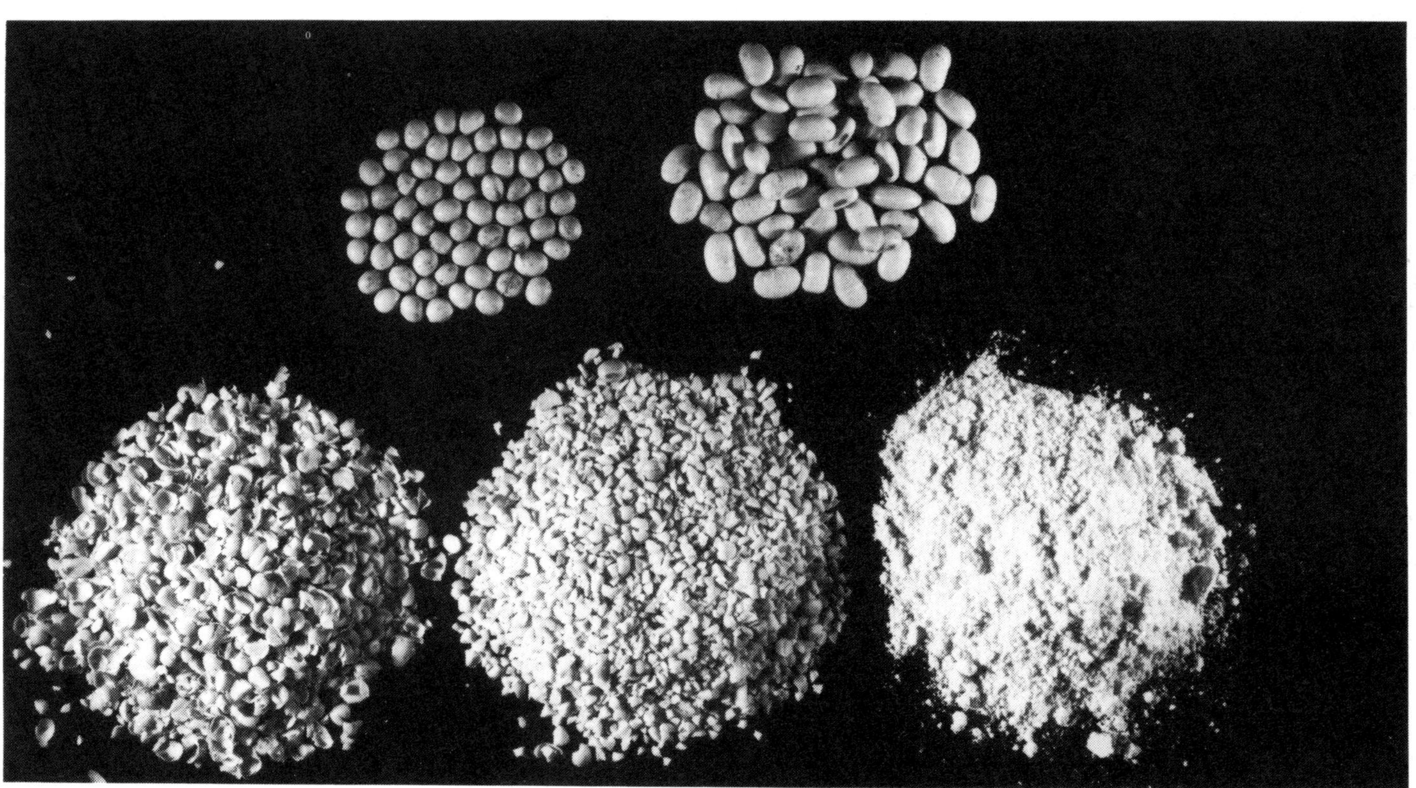

Above   *The soya bean has long been an essential part of the Oriental diet.*

because it bleaches the yellow colour pigment in wheat, and makes the bread whiter. It also improves the texture and helps the bread keep longer.

Defatted flakes left after the oil has been extracted are the basis for most of the food uses of soya. First the flakes are heat treated to remove any remaining traces of solvent from the oil extraction process. This must be done carefully to ensure that none of the soluble protein is lost. The flakes are then deodorized to remove any unpleasant taste and some are ground to produce defatted flour.

Soya flour is used in a wide range of everyday products including baby foods, biscuits, soups, gravies, stews, sweets, savoury snacks, and slimming and health foods. It is also the basis of textured vegetable protein, which has been called 'tomorrow's food today'.

The defatted flakes are the starting point for two important ingredients of many foods – soya protein concentrate and soya protein isolate.

Soya protein concentrate and isolate

Soya protein concentrate is obtained by removing the soluble fraction of carbohydrate from the defatted flakes. Care has to be taken to ensure that the soluble protein is not washed away during this process. So it is fixed by treating the defatted flakes with alcohol and applying gentle heat. The resulting concentrate, which contains 70 per cent protein, is then spray dried to a fine flour.

Soya concentrate is used in the meat processing industry. Small amounts are added to sausages, frankfurters and luncheon meat to help bind the meat and stop the fat and juices running out when the meat is cooked. The addition of soya concentrate to sausages reduces their tendency to split and spit.

Soya protein isolate is produced by isolating or extracting the protein from the defatted flakes. To do this the defatted flakes are treated with a mild alkali solution which makes the protein soluble. The remaining fibrous material is then separated from the soluble protein. Acid is then added to the soluble protein to make the protein insoluble. It forms curds which are

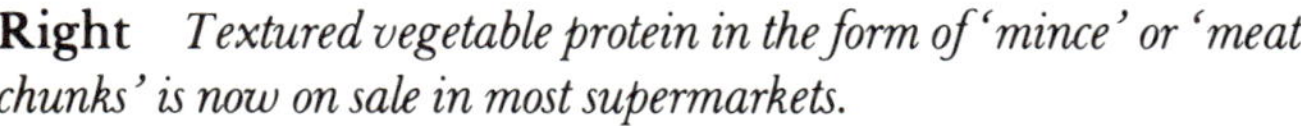

Right *Textured vegetable protein in the form of 'mince' or 'meat chunks' is now on sale in most supermarkets.*

50

separated from the liquid, neutralized, washed and dried to form a 90 per cent protein-rich powder. This process is similar to that used in making cheese.

Soya protein isolate is also used in meat products particularly where foods are processed at high temperatures. It is also used in fish and bakery products, in dairy foods such as whipped toppings and frozen desserts, in processed food where a high level of protein is required, and in dietetic and hospital feeding. It has been used for over 20 years as a source of protein in milk-free products for babies. Soya isolate is the starting point for spun textured vegetable protein and for hydrolyzed soya protein which is used to flavour meats.

Above   *Some of the many food products which have soya as an ingredient.*

Textured vegetable protein

Once the exceptionally high protein content of the soya bean had been discovered, much research was conducted on how to incorporate soya beans into the everyday diet. We could, of course, just eat the soya beans, but that would make for a rather monotonous diet. The solution was textured vegetable protein (known as tvp).

There is nothing magic about the manufacture of tvp. It is made by a process similar to that used to make spaghetti and other types of pasta. Soya flour, mixed with water, is cooked in an 'extruder', which looks like a huge kitchen mincer. As the mixture is forced through the machine, ribbons of tvp come through the holes or nozzle at the end. They are then sliced into small pieces and dried. The nozzle and cutter on the extruder can be adjusted to produce chunks, granules and other shapes. The texture can be varied to look like beef, chicken or other foods. Flavouring can also be added during the process, or the tvp can be left unflavoured.

Another method is to spin the isolated soya protein into fibres. The high protein soya isolate is made soluble again, and the solution is then

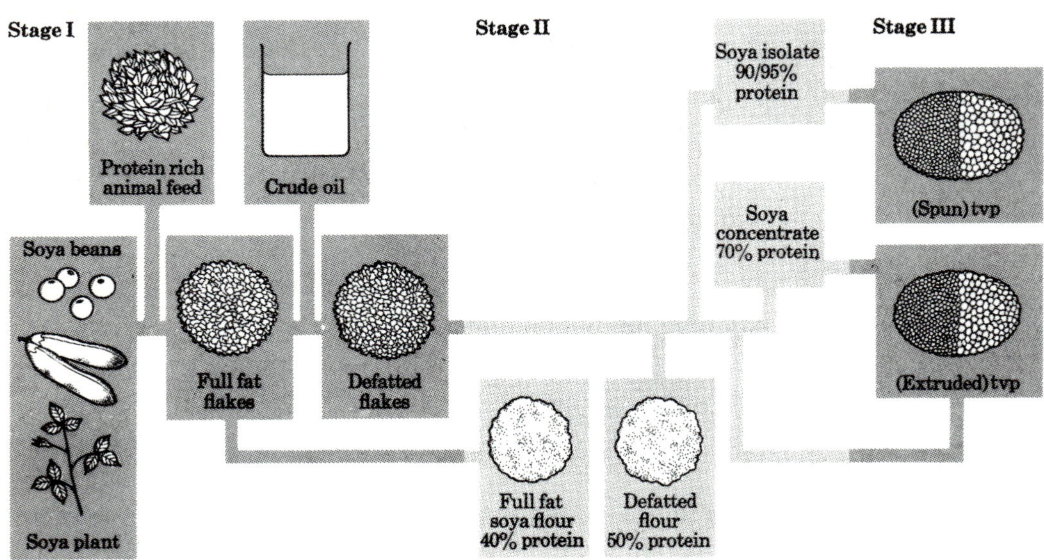

Left *Diagram showing the stages in the processing of tvp.*

Above *Tvp products have the advantage that they keep longer than fresh meat.*

forced through a platinum disc with thousands of tiny holes, called a spinneret. As the fine strands come through they pass into a fixing bath. Bundles of the fibres are then stretched up to four times their original length. The more they are stretched the chewier they become, so the process must be stopped at just the right point to get the right texture. The bundles of fibres are then flavoured, coloured and shaped before packing.

A wide variety of vegetable protein products can now be bought in shops and supermarkets. Some brands of tvp can be used on their own, or they can be used as extenders to supplement parts of the meat content of a dish. Tvp is so versatile that it can be used in dishes as varied as chilli con carne, quiche, rum truffles and mince pies. It is now used extensively in school meals, and other forms of large-scale catering.

In February 1976 the Consumers' Association's magazine *Which* reported that 'textured vegetable protein is a very useful food … and even at today's prices, you'll save money by eating it, instead of meat, especially if you use it on its own …'

An independent survey of housewives in 1974 showed that burgers with tvp were preferred to all meat ones. The winner of the 'Sausage Roll of the Year' competition has used tvp in his winning sausage roll!

Above   *Coatings for paper products, ceiling tiles, paint, petrol, rubber heels and plywood are just some of the products in which soya is used.*

54

Industrial uses of soya

So far we have only considered the ways in which soya is used in the manufacture of food for human and animal consumption. But soya is also used in hundreds of industrial processes. These range from the manufacture of adhesives and plastics to dynamite and waterproofing preparations.

For many years soya bean oil has been used in the manufacture of paints, both for domestic and industrial use. It is also used in the printing industry in printing inks, adhesives and paper. In the U.S.A. the soya bean is used in the brewing industry to help the yeast ferment and enhance the flavour of the beer. The remarkable adhesive power of soya glue has been used to full advantage in the manufacture of plywood. Other products where refined soya oil is used include candles, disinfectants, enamels, fuel and lighting oil, insecticides, leather, linoleum, soaps and paper coatings.

Streptomycin, which is an important germ killer, is made by growing cultures in broth containing soya meal. This meal is also used in making insect sprays, fertilizer, linoleum backing, paste and powder paints, and sizing for cloth.

Soya is even used in fire fighting as a foaming agent. When mixed with water and air it produces fire-fighting foam. Water might put out the surface flames, but the fire could still be burning underneath. When the water drains off, it could start up again. Fire-fighting foam clings to most surfaces and does not drain as quickly as water. Foam is especially useful in putting out liquid fires. If oil has caught alight, the foam floats on the surface and smothers the flames.

Glycerine, which is a by-product of vegetable oils including soya, is used in nearly every industry. It is an ingredient of many medicines including cough mixture and is also a basic medium for toothpaste. Glycerine is sprayed on tobacco to stop it from drying out during processing. It is also used in glues to stop them drying out too quickly. Nitroglycerine is used in the manufacture of explosives such as dynamite.

The wide variety of industrial applications for soya emphasizes the amazing properties of this humble bean. No doubt as research and experiments continue, yet more uses for this versatile crop will be found.

Above   *China is one of the world's major soya bean-producing countries.*

Soya as a world resource

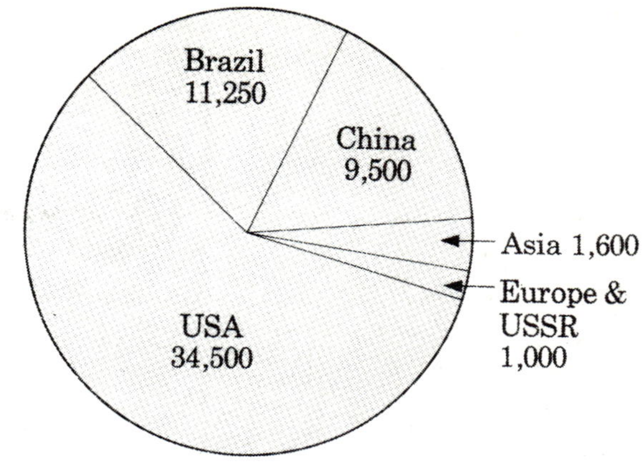

Above *A diagram showing the proportion of soya grown by the world's major producing countries.*

Where is soya grown?

The U.S.A. is the world's major producer of soya, providing 60 per cent of the world's increasing requirement. The rate of growth in soya bean production in the U.S.A. has been phenomenal. In 1924 just over half a million acres of land was devoted to soya bean cultivation, but by the late 1970s this figure had risen to over 70 million acres. Today there are over 530,000 farmers growing soya in 26 major producing states. The top five are the so-called 'Corn Belt' states of Illinois, Iowa, Missouri, Minnesota, and Indiana. These five states produce well over half the total U.S. soya bean crop, which is now the country's largest crop in terms of area harvested.

There are a number of reasons why American farmers like growing soya, the main reason being the high price soya has fetched on the world market in recent years. More processing plants have been built, producing more soya that can be sold overseas. American farmers have been encouraged to grow more soya as the export market has expanded and demand at home has increased. Soya is also well suited to mechanization. Planting and harvesting can be done by machines, and new chemicals help control weeds and pests. All this means higher yields which can be exported and bring money into the country.

For the first half of the twentieth century China was the major soya bean producer, but by the 1950s the U.S.A. had overtaken China

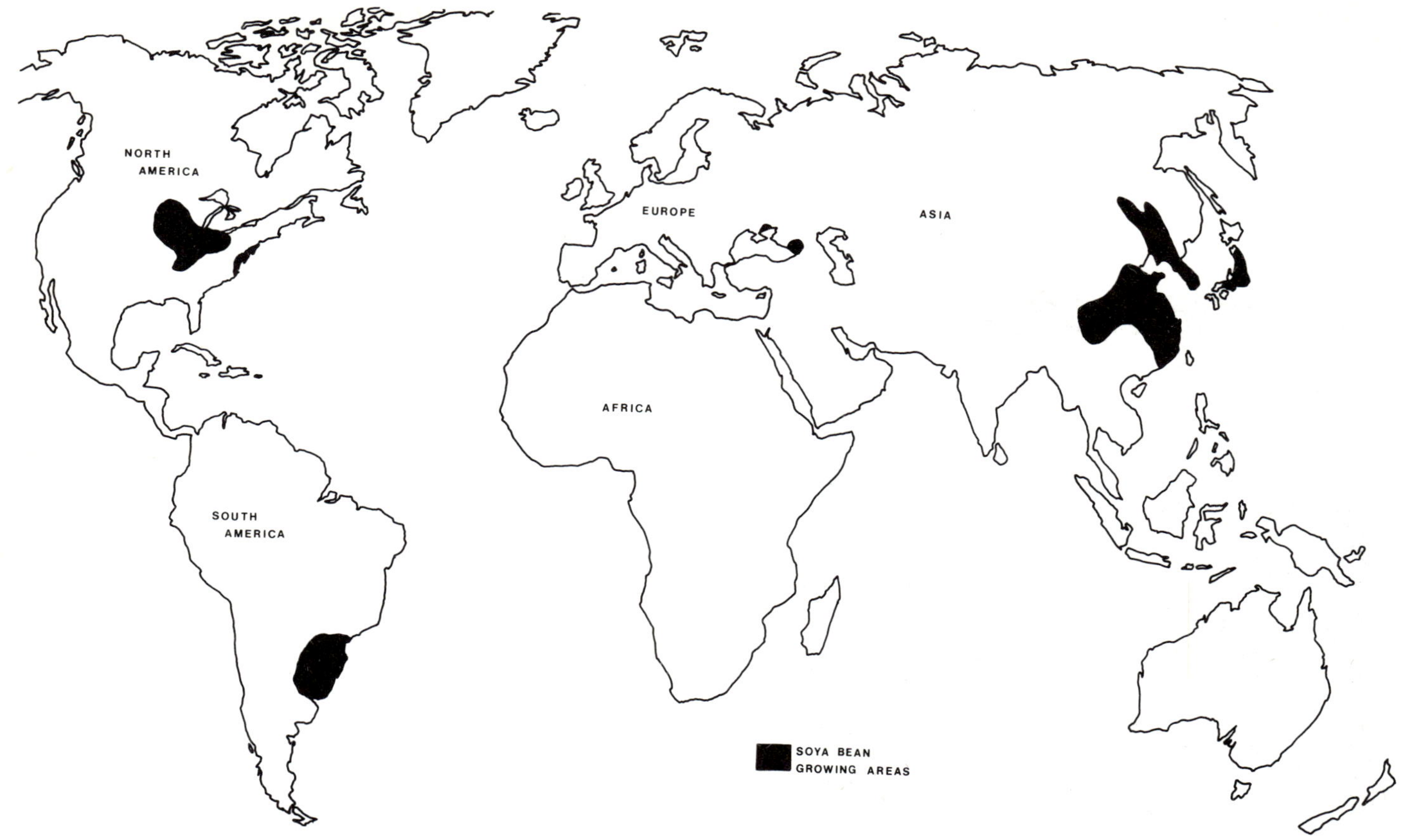

Above *Map to show the main areas of the world where soya is grown.*

and then the entire Orient in soya bean production. China now produces only one sixth of the total world supply, most of which is needed to feed the country's large population. Soya is also grown in Indonesia, Thailand, Kampuchea, Japan, Korea, India and the Philippines. In all these countries soya is a staple food.

Brazil is also a major producer, second only to the U.S.A. In 1980 Brazil's crop is expected to total over 15 million tonnes compared with 10 million the previous year. Rio Grande do Sul

58

and Parana are the main soya raising states in Brazil, but new tracts of land are continually being opened up. Over one fifth of Brazil is covered with grass scrubland, which is being cleared for agriculture, and makes ideal land for soya bean cultivation. Soya has taken the place of some crops formerly grown in Brazil such as maize and cotton. Other Latin American countries where smaller quantities of soya are grown include Mexico, Colombia, Paraguay and Argentina.

Soya is also grown in small quantities in some parts of Europe with warm summers, including southern France, Italy, Spain, Yugoslavia, Hungary, Bulgaria and Russia. Many African countries such as Nigeria, South Africa, Tanzania, Kenya, Zambia and Zaire have also started growing soya.

Above *Zaire is one African country where recent attempts to grow soya have proved successful.*

The world's marketplace

The U.S.A., Brazil and other major soya exporting countries all compete with each other to obtain the best price for their soya on the world market. Soya has to compete with other products such as fishmeal from Peru, sunflower seed oil from Russia, rapeseed from Canada, coconut oil from the Philippines, peanuts from Africa and palm oil from Asia and Africa.

In 1972 soya overtook maize to become the number one cash crop in the U.S.A. This means that the total value of the soya crop was greater than that of wheat, maize, cotton or any other crop. Soya still maintains this position in the U.S.A. Much of the American crop is sent overseas. In 1979 the value of the soya bean crop

Below *Soya is an important item of trade on the world markets.*

Above   *Over half the American soya bean crop is destined for the export market.*

was over 14 billion dollars, nearly 8 billion dollars' worth of which were exported.

Soya plays a vital role in the economy of the U.S.A. Because foreign countries buy so much soya, the U.S.A. has more foreign money to buy other products from overseas.

Soya bean meal is the most important protein meal in the world, accounting for well over half the world supply of meal. Soya oil is an important edible oil, accounting for about one fifth of the world edible oil supply.

Japan is the largest international importer of soya beans, and Europe also takes large shipments. Soya bean oil is exported to India, Pakistan and Mediterranean countries where it is used as a substitute for olive oil.

The future of soya

Man has experienced many changes since the days when people lived in simple rural communities taking their food straight from the land. Today most people in the developed world live together in towns and cities, having little contact with the land. New ways of distributing food and making it last longer have been found. The advent of supermarkets, canning, preserving and freezing has given rise to the 'convenience revolution'. Pre-packed, easy-to-prepare foods are in increasing demand as more housewives and mothers go out to work. Ease of travel and holidays abroad have introduced us to new and foreign foods. In our homes new furnishings, household equipment and products have made our lives more comfortable.

Soya fits well into this pattern of change. Already our lives are touched by soya, whether it is in the linoleum we walk on, the print on the newspapers we read, or the food we eat.

Right *As the cultivation of soya becomes increasingly mechanized, higher yields will be obtained.*

Opposite *This farmer in Zaire is roasting his soya beans in a crude roaster.*

Above *Soya beans fermenting in giant jars before being made into soya sauce in Hong Kong's largest canning works.*

Above *Soya beans are transported along the River Mississippi to processing plants where they will be made into protein-rich food.*

But soya does not only fit into our technological age. Its future also lies in underdeveloped countries such as parts of Africa. We live in a rapidly expanding world. Pictures of starving people testify to the problem of an increasing population drawing on a finite food supply. Soya grown in large quantities in these countries can provide the local population with an important source of protein. As these countries develop and life becomes more sophisticated, soya will become increasingly important.

As new technology discovers yet more applications for this versatile little bean, there is little doubt that we shall be hearing more about soya in the years to come.

Glossary

Amino acids the components into which proteins are broken down during digestion before being absorbed into the blood.

Annual a plant that completes its life cycle in one year.

Bacteria microscopic organisms, some of them harmful, which are present everywhere, including the human body.

Bran the skin of a seed which is separated from the fleshy part during milling. Bran is thought to be essential for good digestion.

Cash crop a crop which is grown for sale rather than as food for the local population.

Catalyst a substance which triggers off a reaction between other substances.

Crop rotation a system of growing different crops on the same land in successive years to avoid soil exhaustion.

Emulsifier a substance which allows water and oil to mix without separating.

Forage crop crop which is grown for cattle food.

Green manure a crop that is ploughed into the land to enrich the soil.

Hydrogenation a process which makes oils solidify.

Insecticide a substance used to destroy insect pests.

Irrigation supplying the land with water by means of artificial ditches and canals to promote the growth of food crops.

Lecithin a substance found in such foods as egg yolk, wheat and soya beans, which the body uses to build up tissues of the nerves and brain.

Protein the body's main building nutrient essential to all living things.

Protein conversion ratio the amount of food an animal needs to produce a certain quantity of food.

Refining to make something free of impurities. For example, crude oil is refined to make it edible.

Self-fertilization a form of reproduction where plants containing both male and female flowers fertilize each other.

Shortening fat used to make pastry rich and crumbly.

Solvent a substance, usually a liquid, which makes another substance dissolve.

Staple food a crop which forms the main part of the diet of the local population.

Trace elements various chemical elements which are found in very small quantities in organisms and are essential for many bodily functions.

Textured vegetable protein (tvp) high-protein food produced from soya, which can be prepared to resemble meat in appearance, flavour and cooking characteristics.

Vegetarian a person who lives on a meatless diet.

Further information

If you would like to find out more about soya, you may like to read the following books:
G. J. Binding, *About Soya Beans* (Thorsons, 1970)
R. G. Whisker and P. Dixon, *The Soybean Grow and Cook Book* (Duckworth, 1977)
T. and J. Watson, *Vegetable Oils* (Wayland, 1981)

The Vegetable Protein Association, representing leading manufacturers and suppliers, has published a comprehensive teaching pack entitled *Vegetable Protein Foods – tomorrow's food today*. For further information apply to the Vegetable Protein Association, 6 Catherine Street, London WC2B 5JJ.

Readers in the U.S.A. may like to contact the American Soybean Association, 777 Craig Road, P.O. Box 27300, St. Louis, Missouri 63141 for further information about soya.

Index

Picture acknowledgements

The publishers would like to thank the American Soybean Association for their helpful comments and advice during the preparation of the manuscript and for providing most of the photographs.
Other pictures on the following pages were provided by J. Allan Cash 64; Forbes Publications Ltd./Vegetable Protein Association 44, 52; Keystone Press Agency 28, 56; Livestock Farming 43; Oxfam 15, 59, 63; United States Department of Agriculture 17, 25, 36; Wayland Picture Library 8, 9, 49, 50, 53. The map on page 58 was drawn by T. J. Browne.